HURRICANE / DISASTER SURVIVAL & COOKING GUIDE

By Hope (Dennis) Comrie

ISBN 978-0-7414-4319-9

Published by:

INFINITY
PUBLISHING.COM

1094 New DeHaven Street, Suite 100
West Conshohocken, PA 19428-2713
Info@buybooksontheweb.com
www.buybooksontheweb.com
Toll-free (877) BUY BOOK
Local Phone (610) 941-9999
Fax (610) 941-9959

Printed in the United States of America

Published January 2013

DEDICATION

This book is dedicated to the memory of my late grandmother – Francis (Willel) Dennis.

To my son Jermei M. DanHamidu

And

Agatha (Pam) Hylton and Vincent Hylton

ACKNOWLEDGMENTS

I want to acknowledge the following people and say thank you: Agnes and McNolan Pinard, Dr. Maureen Lowery, Thea Paulson, Dr. Alexander Ferreira, Dr. Peter Ferrer, Dr. Dolores Tamer, Dr. Grace Wolff, Dr. Paolo Rusconi, Dr. Ming-Lon Young, the late Dr. Otto Garcia, Dr. Michael Cavanaugh, Dr. Donald Rosenberg, Yvonne Garcia, Diane McMullen, Tess Macamos and Lili Hernandez and more. Thank all of you for helping me thru my fire disaster.

Carthy Thomas for going over my manuscript over and over, Kris Schaffer for proofreading, Professor Frumkin of Broward Community College, my son's professor, for reading and critiquing, Siddy Coakley for encouragement and information, and my son, Jermey, for reading my manuscript over and over again with great feedbacks.

TABLE OF CONTENTS

INTRODUCTION

I moved from New York to South Florida after living in New York for 20 years. I could no longer take the snow storms or the slush when it starts to melt. Some of the subway stations so crowded, we had to wait for a train to come into the station, people to get in the train and clear the platform and those of us standing on the stairs can now enter the platform to wait for the next train. One morning I said, "This has to change, I can no longer do this," but I did not know how and when.

Then there is that car towing they do in New York, if you are parked someplace you had no idea was illegal to park there, or you are late coming downstairs to move your car because you did not have a parking space for the right side of the road for today (they clean the street everyday – one side today, the other side tomorrow), all you see is the back of your car turning the corner hooked up on a tow truck, I would say to myself, "this has to change I am moving someplace where I can park my car and don't worry, because, I must have a car, even if I live in Manhattan."

My son was now 12 years old and I told him we would leave New York City when he got to that age, because, although he was born there, he hated the cold weather; but we were suppose to move to Savannah Georgia, no idea why we moved to South Florida. But I have no regrets; the weather is amazing here, except for the hurricanes or close calls.

In Hurricane Wilma, when the forecasters were sure it was going to hit South Florida, I had already stocked up on water, canned foods, manual can opener and nothing else. I was grossly unprepared, I needed so much more. I needed prescription medication, went to the drugstore three days after the hurricane passed, but they too had no electricity and was closed.

This book should help those of us who could use more information to be better prepared for a hurricane or any disaster, regardless of where we live. You will be able to make a better check list and include things you never thought of. Do not forget to include the kids, your pets, the elderly, or disabled. Preparedness is not only personal, it's community based.

CHAPTER I

THE HURRICANE WILL COME

If you live in the Southern United States, especially the State of Florida, the Caribbean Islands, or the Gulf Coast, take note - Hurricane season is right around the corner, it runs for six months, from June 1st to November 30th, every year. And Disasters can happen anytime, any place.

So you must be prepared for Hurricanes, Disasters, or any Emergencies, not only at home, but school, work, and your community.

Remember, we do have Hurricane warnings, but Disasters strike quickly and without warning. It can force you to evaluate your neighborhood, workplace or school or can confine you to your home. What would you do if basic services – water, gas, electricity or telephone—were cut off? Local officials and relief workers will be on the scene after a Hurricane/Disaster, but they cannot reach everyone right away or at the same time. Therefore, the best way to make you and your family safer is to be prepared before a hurricane, a storm or disaster strike.

CHAPTER 2

PREPARATION IS KEY
HOW TO PREPARE

THINGS YOU SHOULD DO TO PREPARE:

The off season is the best time to prepare, like NOW.

- Don't wait until the day before the hurricane to shop.
- If you wait until the hurricane is a day or two away the chances of finding the items you will need are slim to none.
- Listen to the weather forecasts and check your emergency supplies.
- Start shopping early, we live in a hurricane zone and there will be hurricanes or close calls.
- During the non-hurricane months, get one or more large durable plastic container/s.
- Place the container in a secret place and shop for canned and other food products early.
- This will be your Hurricane/Disaster Survival Kit (HDSK).
- Place the food items in the container/s and do not allow your family to use them.

- Don't wait for the Local, State or Federal governments to act before 72 hours after a hurricane.

- Prepare for yourself and your family to be self-sustaining at least one week after a hurricane.

- You should have enough food and water to last for two weeks.

- You don't have to shop for everything at once, every week when you go shopping, pick up extra.

- Don't go unprepared then park you Mercedes, Lexus or BMW outside a Government supplies area to pick up free ice, water, and food kit. Allow them to help the people who can't help themselves, like the shut-ins, the disabled, and the elderly.

- You don't have to be rich to prepare for the hurricane all it takes is a little planning, shop as you go and if you follow the following tip you will be well prepared financially:

  One of the Red Cross' tip of the day suggest that you start saving for a rainy day. Stash away a little cash for a true emergency – keep it in your hurricane/disaster survival kit. Again, this doesn't need to be an overwhelming task. We pay bills every month, consider your emergency funds among your debts and put a few dollars away for the future. For example, every time you get paid put away $20.00. Remember, this is not a savings account that you can dip into and you'll want to keep it safe and secure at home so it is easy for you to access when you need it.

And Remember
Take the time to prepare your kit, once you have done it you don't have to think about it.

CHAPTER 3

YOUR SHOPPING LIST INCLUDING MUST-HAVES, AND MUST DO's

WATER

You can live without food for a few days but you cannot live without water. You don't have to run out and buy bottled water, you can drink tap water, remember we live in the USA and we work hard at water quality. Begin to fill up all the empty plastic gallon bottles you have around the house, be sure you put the caps back on and place them in a cool/dry place. Depending on the size of your family you will need 2-one-gallon bottles per adult per day. Then, before the hurricane strikes, fill up all available containers around the house, including your bath tub.

Recycled milk crates can be used as carriers for your water supply. Always keep your water portable. Remember the kids and the elderly also need easy access to water so have smaller bottles for them and you can refill from the larger bottles.

It is a good idea to place as many bottles of water in the freezer as it can hold. A freezer is more efficient when it's full. If power goes out, the frozen bottles will keep everything cold and you will have cold water for a few days and things will stay frozen longer.

If you use only bottled water and you go to the grocery store and the water shelves are empty, Frank Calderon of Miami Dade County writes:

Bottling Your Own Water is Safe and Easy

Instead of buying bottled water, invest in plastic water containers for your family. Plastic water containers are available in a variety of sizes, from four to 10 gallons or more, and some are collapsible and can be folded easily for storage. They can be purchased at either sporting goods stores or the sporting goods department of most major department stores. And they are built to last for years of use, so you'll save money in the long run.

Make sure you get enough plastic containers so everyone in your family – including your pets – has enough water to last a minimum of three days. Calculate about one gallon per person, per day and you should be okay.

When you bring your plastic containers home, don't fill them up just yet. Keeping water stored a long time could attract harmful bacteria and make the water taste stale. Wait until a hurricane warning is announced first.

Worried that the container might be "icky" when you fill it? Not a problem. Wash it out with soap and water first, and then rinse it well. Next, fill it with a solution of one table spoon of unscented household chlorine bleach - the kind used for laundry, (Clorox or its generic brands, my addition) per gallon. Let it sit 10 minutes, then pour out the solution and rinse the container well. It's now ready to be filled up with tap water.

Follow this advice and you won't have to worry that your grocery store is out of bottled water the next time a hurricane heads our way.

MUST-HAVES

- **YOU MUST HAVE YOUR MEDICATION**

  Call in your refill early and pick up early.

- **RAIN PONCHOS**

  For everyone

- **YOU MUST HAVE BABY FOOD/FORMULA/DIAPERS**

  For at least one week

- **YOU MUST HAVE CASH ON HAND.**

  The banks might not be opened.

- **YOU MUST HAVE A SMALL PLASTIC CARRYING CASE TO PUT IMPORTANT DOCUMENTS.**

  A waist pouch will be handy.

- **YOU MUST HAVE MANUAL CAN OPENERS!**

  Not one but several. I found some at the dollar store for, yes, a dollar, buy several.

- **YOU MUST HAVE A GRILL**

  Coal or propane.

- **YOU MUST HAVE COAL FOR THE GRILL**

  Store it on a shelf in case of flooding.

- **YOU MUST HAVE FIRE STARTER OR LIGHTER FLUID FOR THE COAL**

  If you don't have lighter fluid use newspaper and pieces of twigs before you put the coal in.

- **YOU MUST HAVE WATER**

- **YOU MUST HAVE FOOD**

  Your kit should have everything you need.

- **YOU MUST HAVE YOUR PRESCRIBED MEDICATIONS.**

  Your local drugstore might not have electricity.

- **YOU MUST HAVE SOME FORM OF FIRST AID KIT**

  With gauze, band aids, antibiotic ointments, pain medicine, safety pins, antibiotic ointment, alcohol pads, adhesive tape, sunscreen, lubricant, disposable gloves, scissors, etc.

maintains a disaster relief fund specifically for dealing with local disasters. This money is most often used to provide food, clothing, temporary shelter and other special needs to affected families.

Even though fire kills more Americans each year than all natural disasters combined, of those polled, two-thirds said that they thought hurricanes, tornados, floods and mudslides posed a greater threat to their homes than fires.

"Although many residential fires are preventable, fires can ignite quickly and without warning," said Heidi Taylor, fire safety expert at the American Red Cross. "It is important that every family plan ahead and practice their escape from residential fires – don't wait until they happen."

Losing a home to a fire is a traumatic experience for any family, but there are steps you can take all year long to minimize the threat. The Red Cross urges Americans to prepare their families for the unexpected. Develop a family disaster plan and take these steps toward fire prevention and safety.

Simple Steps to Prevent Fires

- Keep all sources of fuel (paper, clothing, bedding and carpet/rugs) at least three feet away from all heat sources.

- Always make sure children have full adult supervision while cooking or in rooms with candles or fires. **Never leave burning candles unattended.**

- Keep matches and lighters out of the reach of children.

- Teach young children to notify an adult when they see matches and lighters and encourage them not to play with fire if curious or bored.

Simple Steps to Make Your Home Safer

- Install a smoke alarm inside and outside each sleeping area in your house on each additional level. Test smoke alarms regularly (once a month) and replace batteries immediately. Smoke alarms become less sensitive over time and should be replaced every 10 years.

- Consider having one or more fire extinguishers in your house and know how to properly operate them. Training can be obtained from your local fire department.

- Consider installing an automatic fire sprinkler system in your home.

- Know and learn two different ways to escape from every room of your home. Determine if you need escape ladders for sleeping area on the second or third floor. Learn how to use them and keep them near a window.

- Select a location outside your home as a meeting point for your family to gather after evacuating the house. Practice your escape plan twice a year, especially with children.

Also check out www.acqablox.com for water in a box that remains uncontaminated for up to 5 years. Coast Guard approved. Their telephone number is 562-693-9599.

These are some of the items you might want to stock up on and include in your Hurricane/Disaster Survival Kit. Copy this page and keep in your purse as a guide.

CANNED MEATS Salmon Tuna Sardines Corned Beef Canned Chicken Spaghetti Ravioli and Stews **BEVERAGES** Boxed or Canned Juices Milk Soup	**CANNED VEGGIES** Red Beans Black Beans Chic Peas Garbanzo Beans Kidney Beans Mixed Vegetables Baked Beans **MILK** Condensed Milk Dried Milk	**CANNED FRUITS** Peaches Pears Fruit Cocktail Mango **FRESH FRUITS** Oranges Grapefruit Apples Green Banana Green Plantains Sweet Potatoes	**BAKED GOODS** Sliced Bread Buns Pita Bread Saltines Cookies **GRAINS** Cereals Crackers Wheat Thins Granola Bars
DISPOSABLE Paper Plates Paper Towels Plastic Forks/ spoons Wet Wipes	**KITCHEN TOOLS** Manual Can Opener Manual Bottle Opener Barbecue Tongs Forks Spatula	**EXTRAS** Instant Coffee Tea Powdered Creamer Sugar Salt Pepper Spices	**DO NOT FORGET** **BABY FOOD** **PET FOOD**

AND TAKE NOTE:

- **YOU MUST KNOW WHERE YOUR IMPORTANT DOCUMENTS ARE:**

  Deed/Lease to the House, Insurance Policy, Birth Certificates, Passports, Diplomas, Leases, tax records, proof of residence, Credit Cards #s etc. These things can be replaced after the storm but why go thru the hassle.

- **YOU MUST KEEP YOURSELF AND YOUR FAMILY SAFE**

  Be careful of going outside after a storm as there may be live electrical wires on the ground, and even more dangerous the wires may be in the water.

- **YOU MUST FIND A SAFE PLACE IN THE HOUSE TO STAY**

  Preferably in a room without windows and the door can be safely closed...

- **YOU SHOULD SAFEGUARD YOUR MEMORIES**

  Place your important photographs and rare items in zip lock and plastic bags to avoid water damage. They cannot be replaced.

- **STAY AWAY FROM WINDOWS AND DOORS.**

  During the storm do not go near the windows or glass doors, stay at a safe distance.

OTHER IMPORTANT ITEMS FOR THE HOME

- Coolers (I heard you can now buy 5 days or 10 days coolers, check out the home supplies stores). Separate food when placing them in cooler, raw meats should go in one cooler and cooked food and other items should go in another.

- You must have a Full Tank of Gas - and check your oil, and coolant you should have extras on hand. Also remember to have at least a 5-gallon gas container handy. You cannot use water bottles! Do you remember the gas lines after Hurricane Wilma in Florida? Don't let that happen to you.

- Portable Radio - (portable or stationary - more than one). Look for one with a hand-crank charger. The National Oceanic and Atmospheric Administration National Weather Service (NOAA) sell on.

- Flashlights – more than one.
  Try getting a light source with multiple charging options (solar, or hand crank).

- Moist Towelettes for Sanitation

- A Whistle (Go outside with one so your family and neighbors can hear you in case of an emergency).

- A Battery Powered Clock

- Extra Batteries
  Different sizes for different items.

- One Central Lantern or Lamp for the Living room (There are lots of them on the market, get one now).

- A $5.00 (five dollars) Phone from Walgreen's. (When the lights are out, your electronic phone will not work).

- A Cell Phone

  (If you don't already have one and you don't use one often, buy a prepaid and a card to go with it. Radio Shack, Kmart and Wal-Mart sell them. Some are for less than $30. Learn how to use it; in an emergency you can call someone.)

- Baby Diapers

- Sanitary Napkins/Depends

Remember most of these items once you buy them you don't have to every year. Keep them in a safe place and get them when you need them.

CHAPTER 4

THE MOST COMMON DISASTER IN THE UNITED STATES AND STEPS YOU CAN TAKE TO SAFEGUARD AGAINST IT

Home Fires Are the Most Common Disasters in the U.S.

By Katie Lawson, Staff Writer, Redcross.org (February 16, 2006)

Home fires are the most common disasters in the United States, yet 80 percent of American is unaware that their homes could be at high risk. With heating costs on the rise, alternative heating methods are becoming more common and reliable but few realize that they can often be more hazardous.

A new poll conducted by the American Red Cross found that four in 10 families (42 percent) indicate that they are using alternative heating methods such as space heaters, wood/coal stoves and fireplaces to keep warm this winter. Among families with children under the age of 18, the figure rose to 49 percent. The National Fire Protection Agency reports that children are twice as likely as adults to die in residential fires.

According to the U.S. Fire Administration, alternative heating sources play a part in 74 percent of fire-related deaths. Last year alone, the Red Cross helped families affected by nearly 67,000 house and apartment fires, providing immediate emergency assistance. Each chapter

Safety Tips in the Event that Fire Strikes Your Home

- Once you are out of your home, stay out! Use a neighbor's phone to call the fire department.
- If you see smoke or fire in your escape route, use your second way out. If you have to get through smoke, stay low and crawl toward the exit. If exiting through a closed door, be sure to feel the door prior to opening it. If it is warm or hot, use your second way out.
- If smoke, heat, or flames block your exit routes, stay in the room with the door closed. Signal for help by tying a brightly-colored cloth at the window. Call the fire department to tell them your location if there is a phone in the room.
- For more information and materials visit the RedCross.org or contact your local Red Cross chapter.

Fire-Proof your Kitchen/by W magazine

1. Never leave a burner on unattended.
2. Always turn pot handles in.
3. Rollup your sleeves, dangling cuffs can catch fire.
4. Keep curtains and potholders away from the stove.
5. Watch out for electrical outlet overload.

6. Keep your appliances clean. Buildup grease can catch fire.

7. Be sure your fire extinguisher is full, near the stove and works on grease fires.

8. Test your smoke detector regularly. If cooking smoke sets it off, don't remove the batteries; if you must, replace them immediately after cooking.

CHAPTER 5

MY PERSONAL EXPERIENCE WITH HOME FIRE

My Personal Experience with Home Fire:

On the Thursday before Memorial Day weekend, in the year 2000, I just got off the bus, two blocks from home coming in from work, as I got off the bus I smelled smoke, and I said to myself, "someone's house is burning," as I turned the corner I could see lots of people and emergency vehicles in front of my house, then I said, "my house is burning." Then I froze, I could not move, my son was home, it's way after school.

I stood in the same spot for what seems like forever, scanning the crowd, but did not see my son, my baby, the light of my life. Although I wanted to run, shout, get out of control, asking my neighbors if they say my son, I could not move. My neighbors, who were gathered there, were all looking in the opposite direction for me the direction from which I usually came home from work. On this day I got off a stop before so I entered the block from a different direction.

Then my next door neighbor looked around and saw me, she ran to me and told me there was a fire, and there was also an explosion, I said, "explosion!? WHERE?!!!" She said, "I don't know, Jackie, (her daughter) said she heard it, I wasn't home." I managed to ask for my son Jeremy, she said, "I don't know, the firemen are inside." I died a thousand deaths, I could not walk, I could not breathe, I felt dizzy, but

I had to keep conscious, I have to make sure my son is ok before I faint, so I just sat on the sidewalk where I was standing.

Finally, one of the officers came and explained to me that there was a fire in my house, it's not fully extinguished yet, the firemen are still inside, and then the fire Marshall will investigate how it started. He said my son was home and they wanted to know if he was smoking, why the house caught fire. So I asked him if he knew where my son was and he said yes, he is with a group of officers standing by the police cars, they are questioning him now. "Thank God," I said, "Is he hurt?" I asked, he said, "no, he is just upset, but he is a very intelligent kid, he has given them all pertinent information.

Then I saw my son, Jeremy, it was raining when the fire started, he ran out naked, and he was still naked, in only his underpants. Here it is, the house burned and they suspect my son of arson. I sat in front of my gate with one of the police officers, waiting for the Fire Marshall to emerge from the burned house with the verdict for what seemed like forever. Although I knew my son did not smoke and would do nothing of the sort, I was still nervous, my heart was racing, and all the *what ifs* entered my mind. One of the police officers told me if my son was responsible, I will not see him tonight, I didn't answer.

Finally, the Marshal emerged from the house, his face, hands and clothes were black with smoke, and he looked like a cartoon character. He said to my son, "Yes, you were smoking; you burned your house down!" I was just about to faint, when I heard him said, "I am joking, the house was hit by lightening, an electrical outlet in the master bedroom was not grounded and it got hit by lightening, and tells anybody I said so, I have been doing this for 30 years, and I have not been wrong yet." One of the police officers, completed her report, gave me a copy and left.

I lost everything, I moved from New York going on two years, I was renting the house, how am I going to tell my landlord the house burned. I feel so responsible, then it dawned on me, and I said aloud, "it was hit by lightening, I didn't do anything."

So for the first time I am homeless, all I have is the clothes on my back, no family, just coworkers. One of my coworkers came and got me and I went to her house to sleep, and my son went to stay with one of his playmates. I contacted the Red Cross the next day, and they came and inspected the damage, declared the contents a total loss, gave me $250 to buy clothes for my son and me. I did. I was thankful.

When we finally found a permanent home they gave us money to buy a bed and chest of draw to start out. I considered it a blessing. It was only "things" even if I never get them back; my son and I are ok. Whatever amount of this book is sold, 25% will be donated to the Red Cross so they can go on helping.

A portion of the sale of this book will be donated to the Red Cross to help others.

This was my experience, and as you can see, fire happens even when you are not responsible.

CHAPTER 6

HOW TO PREPARE YOUR HOME FOR A HURRICANE

Guides from the, *Hurricane Preparedness,* website visit them for more information.

PREPARE YOUR HOME FOR A HURRICANE:

- **High Winds** – hurricane-force winds can easily destroy poorly constructed buildings and mobile homes. Debris such as signs, roofing material, and small items left outside become flying missiles in hurricanes. Extensive damage to trees, towers and underground utility lines (from uprooted trees), and fallen poles cause considerable disruption.

- **Most mobile / manufactured homes** are not built to withstand hurricane force winds. Residents of homes not meeting that level of safety should relocate to a nearby safer structure once local officials issue a hurricane evacuation order for their community.

- **High-rise buildings** are also vulnerable to hurricane-force winds, particularly at the higher levels since wind speed tends to increase with height. Recent research suggests you should stay below the tenth floor, but still above any floors at risk for flooding.

- It is not uncommon for high-rise buildings to suffer a great deal of damage due to windows being blown out. Consequently the areas around these buildings can be very dangerous.

- **Protect all windows** by installing commercial shutters or prepared plywood panels.

- **Garage doors** are frequently the first feature in a home to fail. Reinforce them so that they are able to withstand high winds.

- **Outdoors** – before hurricane season, assess your property to ensure that landscaping and trees do not become a wind hazard.

  -trim dead wood and overhanging branches from all trees.

  -certain trees and bushes are vulnerable to high winds and any dead trees near a home is a hazard.

  -consider landscaping materials other than gravel/rock.

CHAPTER 7

TORNADOS FROM HURRICANES

Hurricane can also produce tornadoes that add to the storm's destructive power. Tornadoes are most likely to occur in the right-front quadrant of the hurricane. However, they are also often found elsewhere embedded in the rain bands, well away from the center of the hurricane.

Some hurricanes seem to produce no tornadoes, while others develop multiple ones. Studies have shown that more than half of the land falling hurricanes produce at least one tornado. In general, tornadoes associated with hurricanes are less intense than those that occur in the Great Plains. Nonetheless, the effects of tornadoes, added to the larger area of hurricane-force winds, can produce substantial damage.

We have no way at present to predict exactly which storms will spawn tornadoes or where they will touch down. The new Doppler radar systems have greatly improved the forecasters' warning capability, but the technology usually provides lead times from only a few minutes to up to 30 minutes. Consequently, **Preparedness is critical**.

- When associated with hurricanes, tornadoes are not usually accompanied by hail or a lot of lightning, clues that citizens in other parts of the country watch for.

- Tornado production can occur for days after landfall when the tropical cyclone remnants maintain an identifiable low pressure circulation.

- Tornados can develop at any time of the day or night during landfall. However, by 12 hours after landfall, tornadoes tend to occur mainly during daytime hours.

- When a tornado warning is issued, or a tornado is imminent, move to a small interior room away from windows.

- As a last resort, get under heavy furniture, away from windows.

- If you live in a mobile or manufactured home park, get together with other residents and the park owner/manager to designate safe shelter areas in the park or community.

CHAPTER 8

REMEMBER THE CHILDREN
Hurricane/Disaster Stress on Children

HURRICANE Stress and Children

From The Jackson Health Systems (Jackson Memorial Hospital)

Stress Reactions: Stress reactions are a natural response both before and after a hurricane. The ability of children and teens to cope with a disaster may vary depending upon their age. However, children who have been sufficiently prepared and whose parents handle the disaster well, generally will have less severe and more temporary reactions.

Hurricane Readiness: Talk with your children prior to hurricane season. Teach them some basic concepts about weather and its effects in terms they can understand. Use the TV news or pictures in magazines as teaching tools. Explain that a hurricane is a difficult situation but that it can be handled. Never dismiss a question.

Prepare children for the reality of a hurricane. Just as you would have a fire drill, have a similar drill for a hurricane. Make preparation plans which will be readily accepted.

- Encourage children to listen and follow instructions.
- Make plans of what you would do if a hurricane is eminent.

- Go through steps for safety, show them the most protected areas of the house to wait out the storm, etc.

- Have them make a list of the important items they want to pack for safekeeping.

- Participating of children prior to the onset of a hurricane makes them feel less vulnerable and more confident.

- Let them help you assemble an emergency kit, pack your important items or shop for hurricane provisions.

- Have them pack their favorite clothes and toys in plastic bags to protect from water damage.

- Give them their own flashlight.

Coping Techniques for Children: Children naturally look to parents for reassurance; the more quickly it comes, the faster the emotional wounds heal. Recognize that small children may not be able to verbalize their fears and anxieties. When they are afraid, they are most fearful of being left alone. So include them in your activities following the hurricane. Do not leave them alone in an evacuation center while you go back to tend to the damage. This will help alleviate a "clinging" behavior. Plus, when they see you coping, they will adapt too.

Encourage children to talk. Provide an atmosphere of acceptance in which they feel free to express their anxieties. Include family, friends or other children in the discussions. Allow children to express themselves without fear of judgment or indication of expectations. Another way of talking, but silently is to have children draw pictures of the events surrounding the hurricane. Ask them what the pictures are.

Remember, these reactions and symptoms are normal after a disaster. Usually they will disappear slowly, without psychological assistance. If they last longer than two weeks or they are very intense, where a child seems to be having extended difficulties in handling everyday functions, professional help should be arranged. If you are financially unable to acquire professional help, crisis counseling is provided free of charge to disaster victims through the Departments of Health and Rehabilitative Services (HRS). Contact them in your area.

CHAPTER 9

FOR PEOPLE WITH SPECIAL NEEDS

Establish a Personal Support Network which is made up of individuals who will check with you in an emergency to ensure you are OK and to give assistance if needed. This network can consist of friends, roommates, family members, coworkers and neighbors.

Personal Assistance Services, or attendants, may not be available after a disaster. Therefore, it is vital that your personal support network consist of different people than those who usually assist you.

Do not depend on any one person, work out support relationships with several individuals. It's best to identify a minimum of three people at each location where you regularly spend a significant part of your week; work, home, school, volunteer site, etc.

Seven important items to ***discuss, give to*** and ***practice*** with your Personal Support Network.

1. Make arrangements prior to an emergency for your support network to immediately check on you after a hurricane/disaster and, if needed offer assistance.

2. Exchange important keys.

3. Show your support team where you keep emergency supplies.

4. Share copies of your relevant emergency documentations, evacuation plans and emergency health information card.

5. Agree and practice a communication system regarding how to contact each other in an emergency. Do not count on the telephone working.

6. You and your personal support network should always notify each other when you are going out of town and when you will return.

7. The relationship should be mutual. Learn about each other's needs and how to help each other in an emergency. You could be responsible for food supplies and preparation, organizing neighborhood watch meetings, interpreting.

Equipment and Assistive Devices

You should keep important equipment and assistive devices in a consistent, convenient and secured place, so you can quickly and easily locate them after the disaster.

Make sure these items such as teeth, hearing aids, prosthesis, mobility aid, cane, crutches, walker, respirator, service animal harness, augmentative communication device or electronic communicator, artificial larynx, wheelchair, sanitary aids, batteries, eye glasses, contact lens, including cleaning solutions, etc., are secured.

For example: keep hearing aid, eye glasses, dentures, etc., in a container by your bedside which is attached to night stand or bed post using string or Velcro, oxygen tank attached to the wall, wheelchair locked and close to the bed. This helps prevent them from falling, flying or rolling away during a disaster.

If you use a laptop computer as a means of communication, consider purchasing a power converter. A power converter allows most laptops (12 volts or less) to run from a cigarette lighter on the dashboard of a vehicle. Visit The American Red Cross website for more information.

CHAPTER 10

GUIDES AND COPING TECHNIQUES FOR PETS

For millions of us Americans, pets are an integral part of our lives. Whatever emergencies we face, they too should be included.

TAKING CARE OF YOUR PETS' NEEDS

Before the disaster:

- Make sure that your pets are current on their vaccinations.
- Pet shelters require proof of vaccines.
- Have a current photograph.
- Keep a collar with identification on your pet and have a leash on hand to control your pet.
- Have a properly-sized pet carrier for each animal – carriers should be large enough for the animal to stand and turn around.
- Plan your evacuation strategy and don't forget your pet!

- Specialized pet shelters, animal control shelters, veterinary clinics and friends and relatives out of harm's way are all potential refuge for your pet during a disaster.

During the disaster:

- Animals brought to a pet shelter are required to have: Proper identification collar and rabies tag, proper identification on all belongings, a supply of food, water and food bowls, any necessary medications, specific instructions, and news papers or trash bags for clean-up.

- Bring pets indoor well in advance of a storm - reassure them and remain calm.

- Pet shelters will be filled on first come, first served basis. Call ahead and determine availability.

After the Disaster:

- Walk pets on a leash until they become re-oriented to their home - often familiar scents and landmarks may be altered and pets could easily be confused and become lost.

- Also downed power lines, reptiles brought in with high water and debris can all pose a threat for animals after a disaster.

- If pets cannot be found after a disaster, contact the local animal control office to find out where lost animals can be recovered.

- Bring along a picture of your pet if possible.
- After a disaster animals can become aggressive or defensive – monitor their behavior.
- Don't forget your pet when preparing a family disaster plan.

Pet Disaster Supply Kit:

-Proper identification including immunization records.

-Ample supply of food and water

-A carrier or cage

- Medications

-Muzzle, collar and leash.

-A picture of your pet.

-Old newspapers for sanitation.

Remember after Katrina - so many pets were stranded and so many could not be identified, don't let that happen to your pet, follow the above suggestions and your pet should be safe.

CHAPTER 11

SURVIVAL RECIPES

Although we had a hurricane/disaster we still have to eat and feed our families. The following recipes should assist you in the preparation of year meals from your HDSK (Hurricane Disaster Survival Kit.

NOW LET'S PREP THOSE CANNED FOODS AND GRAINS IN STYLE

Add raisins and fresh fruits – like apples pears, peaches, and banana to cereals.

*

If fresh herbs are available use them for your salads:
Try fresh cilantro, parsley, thyme,
leeks, red onions, and green onions.

*

And to your vegetable salads, add raisins, seeds and grains. Trail mix is also great in salads.

BREAKFAST SUGGESTIONS

NO COOK BREAKFASTS

TRADE WINDS OATMEAL	
2 cups instant oatmeal	Directions: Place all ingredients in a medium bowl except marshmallows and banana. Let stand at room temp. for 15 minutes. Serve with banana and marshmallows.
½ cup plantain chips	
½ cup marshmallows	
1 cup milk	
1 large banana (sliced)	

FRUIT AN' NUT OATMEAL	
2 cups instant oatmeal	Directions: In a medium bowl place all ingredients and let stand at room temperature for about 20 minutes. Serve in cereal bowls. Serves 4.
½ cup slivered almonds	
½ cup raisins	
1 cup milk	
½ cup dried pineapple	
½ cup trail mix	

HURRICANE OATMEAL	
2 cups instant oatmeal	Directions: Mix all ingredients in a bowl. Let stand at room temperature for 10 minutes then serve. Serves 4.
½ cup raisins	
½ cup dried pineapple	
½ cup whole cranberry sauce	
I cup evaporated milk	

NUTTY OATMEAL

2 cups instant oatmeal	Directions: Combine all ingredients except banana in cereal bowls. Let it sit for 15 minutes. Serves 4.
½ cup slivered almonds	
½ cup raisins	
2 tablespoons brown sugar	
1 ripe banana	

FRUIT BREAKFAST

1 banana	Directions: Peel fruits, cut in bite sized pieces, place in an appropriate sized bowl, sprinkle with lemon juice (so fruits will not get brown). Add honey and wheat germ. Serve at room temperature or chilled.
1 pear	
1 peach	
1 apple	
2 oranges	
2 tablespoons honey	
2 teaspoons wheat germ	
Juice of half of lemon	

You get the idea – make up your own emergency breakfast to suit your taste by combining some of the ingredients here and your own. Be creative.

BEAT THE HIGH COST OF CEREAL
MAKE YOUR OWN

NO COOK TEENAGERS CEREAL (Make your own #1)

2 pounds quick cooking rolled oats, crushed	Directions: Combine all ingredients except the honey, mix well. Store in covered jar. Makes 24 servings. Add honey when serving
½ pound sunflower seed kernels	
½ pound wheat germ	
¼ pound almonds crushed	
½ roasted cashews	
½ pound raisins	
½ pound chopped dates	
½ cup dried apricots	
Honey	

NO COOK TEENAGERS CEREAL (Make your own #2)

1 cup cashews	Directions: Chop nuts and combine with oats, wheat germ, currants, apricots and brown sugar. Serve with wheat germ. Stored covered. Serves 12.
1 cup blanched almonds	
3 cups quick cook rolled oats	
1 cup wheat germ	
1 cup currants	
1 cup brown sugar	
1 cup dried apricots	

UNCOOKED CEREAL #3

2 cups instant oat meal	Directions: Combine all ingredients and store in a covered jar. Serve with a sprinkle of cinnamon or nutmeg.
½ cup brown sugar	
1 cup roasted peanuts	
1 cup raisins	

When making Oatmeal on the stove, spray the pot with non stick cooking spray to keep the oatmeal from boiling over or sticking to the pot.

HOT OATMEAL

2 cups quick cooking oatmeal	Directions: Bring water to a boil with salt or cook by package direction. When oatmeal is cooked, add milk, sugar and nutmeg. Serve in cereal bowls topped with raisins and apple slices.
2 cups water	
½ can evaporated milk	
½ teaspoon salt	
½ cup raisins	
2 red apples (cored and cup in slices)	
Sugar/substitute	
½ teaspoon nutmeg	

MILLET BREAKFAST CEREAL

1 cup whole millet	Directions: Place millet, water, milk and salt in a saucepan. Bring to a boil. Cover and let simmer for about 45 minutes. Serve with honey and half and half. Serves 4.
2 cups water	
2 cups whole milk	
1 teaspoon salt	
1 cup cored and sliced apple	
1 cup raisins	
Honey and half and half	

COUNTRY POT HOMINY

2 cups hominy corn	Directions: Soak corn overnight in 4 cups of water. Add salt and boil on medium heat for 20 minutes. Cook to desired taste. Add all other ingredients. Allow to simmer.
4 cups water	
1 can evaporated milk	
½ cup brown sugar	
1 teaspoon nutmeg	
½ teaspoon cinnamon	

GRANNIE WILLEL ANYTIME MEAL

1 15-oz can sardines in tomato sauce	Directions: In a medium sized skillet, sauté bell pepper and onion in oil. In a large bowl place sardines with liquid and all other ingredients. Stir with wooden spoon, until lumpy. Pour in a well greased pan sprayed w/non stick spray and bake for 45 minutes at 350° F until firm to the touch. Eat hot or cold sliced.
Salt (optional).	
1 dozen large eggs	
8 oz. sour cream	
1 large onion, minced	
2 tablespoons garlic powder	
1 large red pepper	
¼ cup coconut oil	
3 tablespoons cayenne pepper	
Zest of two lemons	

SALMON SURPRISE

1 can salmon	Directions: Drain off water, empty salmon in a medium container. Mince parsley, onion, and red pepper, add to salmon mixture, then add mayo and cayenne pepper to taste. Serves 3.
3 sprigs parsley	
1 small onion	
1 red bell pepper	
½ green bell pepper	
1 tsp. cayenne pepper	
2 Tbs. mayonnaise	

SUPER QUICK SALMON SALAD

1 15-oz. can salmon	Directions: Empty salmon in small bowl. Add all other ingredients and combine until moist. Serve with sandwich bread or in pita bread.
3 tablespoon mayonnaise	
1 teaspoon garlic powder	
1 teaspoon onion powder	
Hot pepper sauce to taste	

SALMON CARIBE

1 15-oz can salmon	Directions: In a medium sized bowl, combine salmon, celery, cucumber, peppers, tomato, then all lemon juice, onion, and mayonnaise. Combine to desired consistency. Serve on lettuce leaves or a salad blend. Serves 4.
2 stalks celery	
1 medium cucumber, minced	
½ red bell pepper	
1 small onion	
1 medium tomato	
Juice of 1 lemon	
¼ cup mayonnaise	
1 green bell pepper	

DUNNS RIVER SALMON SALAD

1 15-oz can salmon, flaked	Directions: In a large bowl combine salmon, scallions and cooked shells. Add dressing, and pepper, toss gently to combine. Cover container and refrigerate for about two hours.
12-ounces cooked pasta	
½ cup chopped scallions	
6 hard boiled eggs	
Thousand Island Dressing	
1 Tbs. cayenne pepper	

Salmon in these recipes can be substituted for any can food during the hurricane. You can use canned chicken breast as well as canned ham or any canned fish of your choice. Also if you have left over chicken, beef, turkey or pork, combine them to make salads to feed your family.

MEDITERREAN SUMMER SALAD

2 cans sardines	Directions: In a bowl, toss tomato, sardines, artichoke hearts, or bamboo shoots. Add dressing to desired consistency, then add croutons. Enjoy with bread or by itself.
1medium can tomatoes	
1 can hearts of palm	
1 cup seasoned croutons	
1 can bamboo shoots or	
1 can artichoke hearts	
Thousand Island Dressing	

MONTEGO SUMMER SALMON SALAD

1 can pink salmon	Directions: Place salmon and tuna in large bowl. Use a wooden spoon to mix well. Add tomato, apple, bell pepper and mayo. Served stuffed in pita bread.
2 cans white tuna	
1 cup cherry tomatoes	
1 green apple, minced	
1 red bell pepper, minced	
½ cup mayonnaise	
2 teaspoon cayenne pepper	

TUNA IN A POUCH

2 cans 6-oz white tuna	Directions: Mix together cucumber, tomato, tuna. Add dressing. Line pita bread with lettuce leaves and fill with tuna mixture. Serves 4.
1 small chopped cucumber	
1 medium chopped tomato	
4 small pita breads	
4 lettuce leaves	
Blue cheese salad dressing	
4 pita bread	

CARIBBEAN TUNA

2 cans tuna	Directions: Mince celery and scallions in small pieces. Combine with tuna, then add mayonnaise, pepper and coconut oil. Serve with bread or crackers. Serves 2.
1 stalk celery	
2 sprigs scallions	
2 Tbs. coconut oil	
2 tsp. mayonnaise	
1 tsp. cayenne pepper	

ARUBA SARDINE SALAD

2 cans sardines	Directions: Drain sardines, place in a bowl and add mince bell pepper, onion and lemon rind. Toss with mayo until well coated. Serve as you wish.
1 small onion	
1 jalapeno pepper	
½ teaspoon lemon rind	
½ cup red bell pepper	
3 tablespoons mayonnaise	

SARDINES AND APPLES

2 cans sardines	Directions: Discard oil in sardines and place in a small bowl. Combine all ingredients. Toss until well coated and serve on sandwich bread.
½ red apple	
½ green apple	
1 teaspoon lemon juice	
2 stalks celery, minced	
3 tablespoons mayo	

SPICY SARDINE SALAD

1 15-oz can sardines in tomato sauce	Directions: With a wooden spoon, flake sardines, then add all Ingredients, mayonnaise last. Let sit for a few minutes before serving. Omit one of the peppers if you wish. Serves 3.
1 stalk celery, minced	
3 springs scallions, minced	
3 tablespoons mayonnaise	
1 hot pepper, discard seeds	
1 teaspoon cayenne pepper	
2 teaspoons hot pepper sauce	

SEABREEZE MACKREL SALAD

1 can mackerel, drained	Directions: Remove bones and flake with fork. Add onion, and all other ingredients. Combine well and let sit for 15 minutes before serving to absorb the spices.
1 can chili	
1 small red onion, minced	
3 tablespoon hot sauce	
3 tablespoons olive oil	
1 large tomato, minced	

MACKREL & GREEN BANANA

1 can mackerel	Directions: Sauté onion, red and green peppers, tomato in cooking oil. Add mackerel and jalapeno. Combine and turn to low for 10 minutes. Cook green banana in boiling water with salt until done. Serve both together, hot.
1 large onion	
Cooking oil	
½ each green and red pepper	
3 springs parsley	
1 medium tomato	
1 jalapeno pepper	
2 lbs. green banana	

THE QUICKEST LUNCH/THE SIMPLIEST DESSERTS

Peanut Butter and Jelly Sandwich	Directions: Spread Jam/Jelly on one slice of bread spread peanut butter on another slice. Place one on top of the other. PB&J Sandwich
Peanut butter	
Strawberry jam/jelly	
Sandwich bread	

The Simplest Dessert of All	Directions: Peel bananas and place in freezer using plastic wrap. When frozen slice or serve whole.
6 ripe bananas	

ARUBA SARDINE SALAD

2 cans sardines	Directions: Drain sardines, place bowl. Add mince bell pepper and onion. Add lemon rind. Toss with low fat mayo until well coated. Serve w/bread.
½ cup red or green bell pepper	
1 small onion	
1 small hot pepper	
½ teaspoon lemon rind	
3 tablespoons mayonnaise	

STALE BREAD SALAD

1 loaf stale bread	Directions: Cut bread into small pieces and place in a salad bowl. Sprinkle with water. Add all remaining ingredients and toss.
2 medium tomatoes, sliced	
1 large red onion	
1 cucumber, peeled/diced	
1/3 cup olive oil	
2 tablespoons vinegar	
1 teaspoon black pepper	
Dash of salt	

STALE BREAD AND SARDINE SALAD

3 cans sardines, drain off oil	Directions: In a medium salad bowl, empty sardines, then add all other ingredients. Toss to combine well. Add other spices to suit your taste.
1 medium red onion, minced	
1 stale bread, cut in pieces	
Hot pepper sauce to taste	
1 medium tomato, minced	
Zest of two limes	
3 tablespoons olive oil or	
3 tablespoons mayonnaise	

BEAN SALADS

TROPICAL BEAN SALAD

1 15 oz can of beans, rinsed	Directions: In a medium bowl mix oil, lemon juice and a dash of salt. Add other ingredients, except lettuce. Add pepper last and let salad sit for ½ hour before serving. Serve on lettuce leaves.
½ cup chopped red onion	
1 medium ripe avocado peeped	
3 springs parsley, minced	
1 cup shredded lettuce	
3 tablespoons olive oil	
1 apple peeled and chopped	
Juice of one lemon	
2 tsp cayenne pepper	

SPICY BEAN SALAD

1 small red onion	Directions: Combine all ingredients in a bowl and toss with coconut oil. Let sit for an hour before serving. Use as a side dish or a main dish.
1 small bottle olives	
2 sprig scallions, minced	
2 hot peppers, chopped.	
1 red bell pepper, minced	
½ green pepper, chopped	
4 tablespoons coconut oil	
1 teaspoon cayenne pepper	
1 can pinto beans	
1 can chick peas	

QUICK BEAN SALAD

1 can garbanzo beans	Directions: Drain water from cans, and combine all ingredients into a salad. Use mayonnaise or olive oil to combine and toss well.
1 can chick peas	
1 red onion,	
1 ripe tomato	
1 red pepper, pepper sauce	

DIPS

ONION MIX DIP

1 pint sour cream	Directions: In a small bowl, pour sour cream, add onion mix, stir. Add paprika. Let sit for 10 minutes before serving. Great for chips and raw veggies.
1 envelope onion soup mix	
1 teaspoon paprika	

SALMON DIP

1 6-oz can salmon	Directions: In a medium container place fish, onion and hot pepper sauce, using a spoon mince, then add sour cream + garlic powder.
1 small onion, minced	
3 teaspoon hot pepper sauce	
½ cup sour cream	
1 tablespoon garlic powder	

YOGURT/SOUR CREAM DIP

8 ounces sour cream	Directions: In a small bowl, pour sour cream, add yogurt stir, add soup mix then paprika and pepper. Let sit before serving. Use with chips.
8 ounces yogurt	
1 teaspoon paprika	
1 envelope onion soup mix	
1 teaspoon cayenne pepper	

ADDITIONAL TIPS:

- Before the hurricane, cook a few dishes, store them in plastic containers and freeze them. When you lose power heat them up on your grill.
- Wash fruits, vegetables and cooking utensils.
- Turn your refrigerator and freezer to coldest settings.
- Buy fresh fruits, un-ripened.
- Remember to pick up crackers, Wheat Thins, Triscuits, and saltines.
- Remember to store your canned foods, can opener, and spices/ingredients together.
- Mayonnaise does not have to be refrigerated if you use a clean spoon every time you use it (take out the amount you need at one time and close the jar).
- Remember if you do not have electricity discard all unused food. Make enough for one meal only.

BAHAMA KAY PINEAPPLE- SUNDAE

Ingredients	Directions
4 tablespoon unsalted butter	Directions: In a medium skillet, melt butter and sugar over medium to high heat. Add pineapple and bananas and cook until bubbly, about 2 minutes. Remove from heat and stir in rum. Scoop 1/3 cup each of coffee and vanilla ice cream in 4 sundae dishes. Pour warm fruit sauce over ice cream.
4 tablespoon brown sugar	
1 can crushed pineapple	
2 large ripe bananas, sliced	
½ cup dark rum	
1 ½ cups coffee ice cream	
1 ½ cups vanilla ice cream	

TROPICAL SUN SALAD

Ingredients	Directions
1 cantaloupe (made into melon balls)	Directions: Combine all ingredients except the ice cream in a large salad bowl. Chill for at least 3 hours. Add ice cream in scoops just before serving. NOTE: Other fruits of your choice can be substituted for fruits here. ENJOY.
6 bananas	
½ pound cherries, pitted and halved	
1 pint strawberries	
4 fresh peaches pitted and sliced	
1 can crushed pineapple (or one whole)	
2 cups seedless grapes	
1 can mandarin orange segments	
1 can papaya nectar	
1 bottle/can guava juice	
1 quart strawberry ice cream	

SOUTH MIAMI FRUIT SALAD

1 large mango, peeled, sliced	Directions: Combine all fruits in a medium bowl. Add guava nectar and chill before serving.
1 medium papaya	
1 large orange, 1 tangerine	
1 can pineapple, drained	
½ cup guava nectar	

ISLAND FRUIT SALAD

1 can fruit cocktail, drained	Directions: Combine all ingredients and serve at room temperature or chill.
½ can condensed milk	
½ cup shredded coconut	
Dash of nut meg	

BUY THESE FOR THE NIGHTS YOU HAVE NO LIGHTS, THE KIDS WILL ENJOY THEM.

HEALTHIEST SNACKS

Granola Bars

Dry cereals – granola, raisin bran, corn flakes, etc.

Seeds –sunflower and pumpkin.

Dried fruits – raisins, prunes, apricot, pineapple, mango

Nuts – peanuts, cashews, almonds

While you still have power or long before the hurricane, experiment with a few of the recipes here, for lunch or dinner, that you will have a good idea of which ones you like best when the light goes out. Then you will be sure to have all the ingredients you will need on hand.

Remember if the food looks spoiled, it is spoiled, if it smells spoiled, it is spoiled, toss it. As the saying goes, when in doubt, throw it out.

Canned foods are loaded with sodium so pour water in the can and wash out some of the salt before you combine your recipes.

I hope this book will be invaluable in your preparation for the Hurricane season or any emergency situation we might face in the coming years. Stay Safe.

When the Hurricane/Disaster has passed, don't get too complacent, always be prepared for any emergency by keeping extra canned, dried foods, grains, and water, in your home, at all times.

CHAPTER 12

RECIPES FOR AFTER THE STORM SAVE TIME, SAVE POWER – COOK FOR MORE THAN ONCE

HERE ARE A FEW RECIPES FOR WHEN THE LIGHTS RETURN

STOVETOP BARBIQUE CHICKEN

1 teaspoon cooking oil	Directions: Heat oil in large skillet. Add onion – cook until softened. Stir in ketchup, water, vinegar, and bring to a boil. Add the chicken pieces. Cover skillet and let cook for 30 minutes. Uncover skillet, turn chicken pieces and cook another 15 minutes until tender.
1 medium onion, chopped	
½ cup ketchup	
½ cup water	
2 tablespoons brown sugar	
1 ½ teaspoons soy sauce	
½ teaspoon chili powder	
1 3-pound broiler, cut up	
2 tablespoon vinegar	

TURKEY BURGERS

1 pound ground turkey	Directions: Combine all ingredients and shape the mixture in 6 patties. Grill or fry until done. Serve on hamburger buns, toasted.
½ cup dried bread crumbs	
3 tablespoons chopped onion	
2 tablespoons ketchup	
1 tablespoon lemon juice	
1 teaspoon soy sauce	
½ teaspoon paprika	
¼ teaspoon hot pepper sauce	
½ teaspoon cayenne pepper	

GRANNIE WILLEL ONE POT MEAL

3 pounds chicken thighs	Directions: In a large Dutch pot, heat oil. When hot brown pork, then chicken on both sides. Set aside when done. Sauté onion and scallions, set aside. Pour off remaining oil and all other ingredients except flour/cornmeal and brown sugar, then add the chicken and pork, stir. Cook on medium for 30 minutes. Use flour/cornmeal to make little dumplings and add to pot, then add sugar, simmer for another 10 minutes. Done
3 pounds boneless pork	
1 large onion	
3 sprigs scallions	
1 large can of Yams	
1 10-oz pack lima beans	
4 large carrots	
1 can coconut milk	
Salt, cayenne pepper, thyme	
1 cup flour, ¾ cup cornmeal	
3 tablespoon brown sugar	
½ cup cooking oil	
1½ cups boiling water	

JAMAICAN RED SNAPPER with PAN FRIED BANANAS

2 pounds red snapper	Directions: In a shallow bowl – season the fish with salt, pepper, and thyme and set aside. Beat eggs with fork, cover fish with eggs, then roll fish in bread crumbs and fry in oil. Set aside. Cut bananas down and across, fry in same pan that you fried the fish until brown on both sides. Serve with chicken. NOTE: You can substitute plantains for bananas.
1 cup plain bread crumbs	
1 teaspoon dried thyme	
1 teaspoon ground allspice (pimento)	
1 teaspoon salt	
2 tablespoon cayenne pepper	
2 eggs	
¼ cup coconut oil	
4 half-ripe bananas	
Lime wedges	

QUICK CURRIED CHICKEN

1 3-pound broiler, cut up	Directions: Season the chicken with salt and pepper. In a heavy skillet with cover, heat the oil, brown the chicken lightly on both Sides. Add the onion, celery and garlic to the skillet and cook them 5 minutes. Stir in the tomato paste and broth. Bring the sauce to a boil, reduce heat, cover skillet and cook until tender. Remove chicken from skillet and set aside to keep warm, remove bay leaf and discard. In a blender, puree the remaining contents of the skillet. Stir in milk and add chicken. Bring the sauce to just a boil. Serve over hot rice or pasta.
Black pepper	
1 teaspoon salt	
3 tablespoon vegetable oil	
1 large onion, minced	
½ cup celery, minced	
3 large cloves of garlic	
3 tablespoon curry powder	
1 medium banana	
1 bay leaf	
1 large apple, diced	
2 teaspoon tomato paste	
1 ½ water/chicken broth	
½ cup milk	

OVEN FRIED CHICKEN

3 lbs chicken legs and thighs	Directions: Place chicken pieces in shallow bowl. Cover with milk. Let chicken soak for 30 minutes or longer (refrigerate if necessary). In a shallow bowl combine bread crumbs, cheese, thyme and pepper, one by one dip the soaked chicken pieces in the breading mixture coating the chicken on all sides. Set the coated chicken pieces in a greased pan, lightly greased with oil. Bake at 375° for 45 minutes.
Milk for soaking the chicken	
½ cup dry bread	
½ cup parmesan cheese	
1 tablespoon thyme	
1 tablespoon cayenne pepper	
¼ cup coconut oil	

CORN BEEF AND CABBAGE

Rinse the corned beef in cold water and then place in a large pot. Cover with water and add one large sliced onion. Bring to a boil and then lower heat to a nice simmer; cook about 3½ hours or until almost tender. Add potatoes, carrots and simmer about 30 minutes and then add cabbage wedges on top. Continue to cook until cabbage is tender crisp. Turn the heat off and allow setting for about an hour before serving. Compliments of The Happy Cooker/Wanda's Country Home

A recipe I must include in this book is from my late friend Delores Estwick (Suaad). I have enjoyed this recipe with her so many Sundays for dinner.

SUAAD'S MUSHROOM SOUP CHICKEN

1 3-lb broiler-fryer, cut up 1 can golden mushroom soup 2 teaspoon sea salt 1 large onion 2 teaspoon garlic powder 2 teaspoon pepper flakes ½ cup coconut milk 3 leaves cilantro 6 whole cloves 1 bay leaf Cooking oil 1 green bell pepper	Directions: Season chicken with salt, pepper and all other spices. Let sit overnight or for about 4 hours. Brown chicken on both sides. Set aside. Sauté onion, green pepper. Add bay leaf and cilantro. Then pour off oil. Return chicken to skillet and all add ingredients except soup and cook until tender. Place soup in a small saucepan, with ½ cup water, stir, and bring to a boil. Then add to chicken. Simmer for about 10 minutes more. Turn off stove until ready to serve Discard bay leaf when done.

MEDITERREAN YOGURT CHICKEN WITH BLACKENED SAVORY CHICKEN BREAST WITH SAUCE

From the *Guilt Free Gourmet in Woman's World Magazine*

6 large chicken breast	Directions: In a non-metal bowl, place chicken, yogurt, onions, cumin, rosemary, salt and pepper to taste. Mix well. Cover and refrigerate 3-4 hours. Preheat oven at 375˚F. Place chicken in shallow baking pan. Bake 20 min. In bowl combine first 3 sauce ingredients, mix well. Add pinch of salt. Remove chicken from oven and drizzle ½ tsp olive oil over each breast. Place pan under broiler and cook 5 minutes until onions begin to blacken and chicken is cooked to desired doneness. Serve immediately with sauce. Garnish with fresh cilantro.
1 cup plain yogurt	
2 large onions, diced	
1 tsp ground cumin	
1 teaspoon dried rosemary	
Salt	
Black pepper	
2 teaspoon olive oil	
Fresh cilantro leaves	
Sauce	
1 cup plain yogurt	
¼ teaspoon ground cumin	
½ small onion, grated	
Salt	

CHICKEN IN COCONUT MILK

6 chicken breast	Directions: Season chicken with next 5 ingredients + salt. Brown chicken in cooking old in a covered skillet. When done set aside. Discard oil. Cover skillet let cook for 15 mins. Then pour coconut milk over chicken and let simmer.
2 tablespoon onion powder	
2 tablespoon garlic powder	
1 teaspoon cumin	
1 teaspoon thyme	
1 teaspoon pepper	
1 can coconut milk	
Salt and Cooking oil	

BEVERAGES

SANGRIA DE ESPANOL

2 bottles red wine	Directions: Stir together wine, apricot nectar, oranges, and lemons in a pitcher. Before serving add plums, grapes and stir in seltzer. Pour sangria over ice in glasses. NOTE: can be prepared a day or two before needed. Cover at room temperature.
3 cups apricot nectar	
2 oranges, thinly sliced	
3 cups apricot nectar	
2 lemons, thinly sliced	
2 plums, and thinly sliced	
10 green grapes, halved	
3 cups plain seltzer water	
Ice cubes for serving.	

PLANTERS PUNCH

1 quart Appleton rum	Directions: In a large container place all ingredients. Stir well with wooden spoon, place in covered container in the refrigerator for two or more days. Serve over ice.
8 ounces pineapple juice	
8 ounces passion fruit juice	
8 ounces fresh lime juice	
Grated nutmeg	
Pineapple slices	
Ice	

LEMONADE A LA CARIB

2 cups fresh lime juice	Directions: In a large pitcher, combine lime juice, sugar, stir, add wine and refrigerate until next day. Garnish with sliced limes. Serve over ice.
4 cups red wine	
1 cup fine sugar	
1 large lime	

When I moved to South Florida, I discovered Café Con Leche, (coffee with milk) the best coffee I have ever had and I am hooked. I concocked this version for the days I don't go to work and unable to buy my usual cup from the Cuban coffee shop. I offered cups to my friends and they told me I did not make it. I had to show them.

INSTANT 'CAFÉ CON LECHE'

Place two teaspoons of instant coffee,
one cinnamon stick and 2 teaspoons sugar in a cup.
In a small pot heat one cup of milk on low temperature.
Add to the coffee cup.
Stir and enjoy.

When you return to regular schedule, don't be a slave to the kitchen. A great time saver is to use a crock pot/slow cooker. They are available at Kmart, Wal-Mart, Brandsmart, and other stores. Prepare all your ingredients overnight, place everything in the crock pot, and turn it on "LOW" before you leave for work. Dinner will be ready when you return. Here is a sample recipe:

CROCK POT/SLOW-COOKER SUPPER

3 pounds skinless chicken thighs	Directions: Place onion and green pepper in bottom of a 3-quart slow cooker. Top with chicken and sausage. In a bowl, combine tomatoes with flour and all the spices, pour over chicken and sausage. Cover, cook on low until tender, about 8 hours. Skim fat, stir remaining mixture and add some more spices to taste. Dinner is ready. Wow!
1 green pepper	
1 can diced tomatoes	
2 tablespoons flour	
½ teaspoon thyme	
1 teaspoon salt	
3 springs parsley	
1 pound of sausage, any kind	
1 large onion	
2 cloves garlic.	

Some optional items you might want to add to the slow cooker with this recipe are: yellow yam, white yam, potatoes, cocoa, dasheen, dumplings or sweet potatoes. Since it's going to be a one-pot-meal, your family will not ask for the rice if they have these root vegetables.

MAKE A TENDER JUICY ROAST CHICKEN

In an 8-quart non aluminum pot, stir together 5 cups hot tap water, 6 tea bags, 1 cup kosher salt and 2/3 cup brown sugar until sugar and salt are dissolved. Add 9 cups cold tap water. Place chicken in pot, neck side down; cover. Refrigerate at least 8 hours. Drain chicken, discarding marinade. Roast as usual for the juiciest, most flavorful chicken ever.

ORIENTAL STEWED CHICKEN

2 Tb. Soy sauce, ½ tsp chili pdr. 2 cloves garlic, one broiler fryer (about 3-4 lbs), cut into pieces, 3 scallions, cut into 1" sections, ½ cup chicken broth or water.
In a large saucepan combine soy sauce, chili powder, and garlic. Add chicken and scallions, stir to coat the chicken well, let it sit for one hour to marinate, turning occasionally. Add the broth, bring to a boil, then reduce the heat to low, simmer, until cooked abut on hour. Serve with rice or noodles. The kids will love this recipe.

STIR-FRY PORK WITH VEGETABLES

1 lb. boneless pork, cut into bite sized pieces	Directions: In a small bowl combine cornstarch, soy sauce and mustard. Add the pork, and toss to coat well. Heat a skillet, add the Oil and heat for about a minute. Add the pork, and stir fry until the sauce begins to thicken. Now all the two packs of Vegetables. Stir to combine. Simmer for another 5 minutes. Serve over rice.
2 tsp oil	
½ tsp dry mustard	
1 cup chicken broth	
1 teaspoon cayenne pepper	
2 ten oz. packages of frozen mixed vegetables	
1 Tbs. cornstarch	

With most women working, coming in exhausted, no family should expect a three course dinner or even one course everyday of the week. Store partially cooked dinners. Make recipes big enough to feed your family for two or more meals. Leftovers can be stored in the refrigerator for several days and can often be frozen for months. Then you can prepare "instant" meals during the week when you come in tired and hungry and can't wait to get supper cooked quickly.

If you already have meat cooked, you can utilize the microwave to bake a potato, usually a medium takes 5 minutes, or a ripe plantain, also takes 5 minutes, with a salad or cooked vegetables dinner will be ready in a flash. The week-end is a great time to cook "in bulk" – dinners for the next week or beyond. Store cooked food in the freezer – in those inexpensive reusable plastic containers available in supermarkets.

CHAPTER 13

MAKE SURE YOU HAVE IT RIGHT

As you must have learned by now Preparation is key. Your Disaster Prevention should include:

- Creating Your Hurricane Disaster Survival kit.
- Developing your Family Plan
- Having a safe place to go if you must.
- Making sure your Home is Secure.
- Make arrangements for your pets.
- Creating A Hurricane/Disaster Survival Kit
  Your kit will be useful whether you leave your home or stay in your home.
- Remember to develop your plan and share it with the rest of the family and your friends.
- Find a safe place in your home to ride out the storms. The room should be without windows, if you cant find a secure space in your home then maybe you should ride it out in a shelter or some place else that you deem safe.
- Long before the hurricane preparations should have been done to make your home more secure and able to withstand strong winds and rain as well as flooding.

- Remember, there are pet shelters where your pet can be safe until after the storm. Make prior arrangements.

- Have a plan that makes you and your family as safe as possible in your own home.

- Fires are also possible during hurricanes, make sure you are careful with grills not to use them inside. And no generators in the garage nor inside the house.

The time to plan for a hurricane or any disaster is NOW not when the storm is fast approaching. When gathering supplies to weather any tropical storm, it is important to plan for ***before, during,*** *and* ***after.*** Your needs will vary depending on the circumstances. To be on the safe side, emergency officials warn, "*Assume the worst, hope for the best.*" You can't be too prepared.

Start with a plan that includes checklist of items you will need, a course of action to take and keep it in a safe, easy to access place. The following should help you compile your checklist:

-If you are required to evacuate do you know how, when, and where you will go?

-If you stay will you have enough supplies to last for days or weeks?

-If you leave and can't get back do you have what you need with you?

-Do you have proof of residence handy?

-Wear protective clothing and sturdy shoes.

-Listen to your battery powdered radio and follow instructions of local officials.

-Lock /safeguard your home before you evacuate.

-Remember – if you are told to evacuate, you should.

-And remember – take your Hurricane / Disaster Survival Kit with you. It should contain everything you will need.

Do this to your computer

And before you leave home secure your computer by doing the following: Unplug, wrap in plastic, move away from window, place on a sturdy device, under a table but not on floor to avoid water damage in case of flooding. Back up data on floppy disk and keep in a safe place.

CHAPTER 14

MAKE PREPARATION IF YOU MUST GO TO A SHELTER

If you live in a low-lying area that is prone to flooding, you should go

HELPFUL SHELTER INFORMATION

In a shelter, a folding beach chair makes an excellent alternative to cots. They fold flat, double as chairs and allow you to store a 16 gallon plastic storage tub (your emergency supplies kit) underneath to save space and secure your belongings. These tubs are available at most stores, (Walmart, Kmart, Walgreens). In a shelter, store you emergency kit underneath and use a comforter as a cover - it hides what's underneath.

Also take note that if you go to a shelter it might be cramped, and unheated which can increase the changes of pneumonia, influenza and colds. Therefore, equip your kits with any vitamins or medications you take to guard against getting sick and to cope with being sick.

If you are a smoker, be aware that smoking will not be allowed in shelters. If getting to an outside smoking area may be difficult for you, consider stocking you evacuation kit with nicotine gum or patches available by prescriptions.

A good idea is to store your smaller items in a carryall with handles and multiple pockets, you can grab and go if you need to get out fast. Also purchase a waist pouch where you can keep your cash, credit cards, drivers' license, insurance and registration on you at all times. It's also a good idea to have copies of them someplace else.

Also if you are leaving your home make sure all the windows and doors are locked, your alarm is turned off if there is no power (it's best to contact your house alarm carrier as to your best course of action). And it's a good idea to exchange phone numbers with neighbors, including cell phone numbers.

Make sure all your children are accounted for, especially the little ones, count them when you get in the car to make sure they are all there.

Don't forget to inform family members not living with you where you will be riding out the storm with contact information.

CHAPTER 15

FINALLY…

- You must remain calm.
- Use only a flashlight for emergency lighting. Never use candles.
- Turn off electrical equipment you were using when the power went out.
- Leave one light turned on so you'll know when your power returns.
- Be prepared with a portable radio-and listen to updated information.
- Stay indoors in your safe room.
- If you are in a high rise avoid the top floors, move to a lower floor if you can.
- Continue to watch local news or listen to the weather forecasts.
- Do not use your house phone in lightening and thunder.
- If your house roof or other parts are coming apart, place a mattress over you.
- Do not run a generator inside a house or garage.

- Remember wherever you are going take your Kit with you, in a shelter nothing will be provided - you have to have your own supplies.

- Don't ignore warnings.

- Preparing adequately not only means protecting yourself and your property in dealing with a Hurricane's aftermath but it also saves lives.

- Do not go outside during the storm, winds can become suddenly very dangerous.

Hurricane officials advise that it is imperative that we create a culture of preparedness in our home, and community, that each household has a personal hurricane plan and each member of the family participates in the planning, even the kids.

After the Storm

You must remember statistics show that more injuries occur shortly after a hurricane than during the hurricane. You must be careful of downed power lines and fallen trees. Driving can be hazardous shortly after a storm due to downed trees, signs and traffic signals so stay in doors until its safe outside. You will be informed as to when it's safe.

Everyone might not have electricity, so share with you neighbors. If you have power, a gas stove or a grill, offer your neighbors hot coffee or tea, or have a community breakfast, anyone with a grill can take turns. You should have had dried milk to mix to give the kids cereal so now is the time to start utilizing your Kit.

<u>You must have all the supplies you will need to be comfortable:</u>

Your kit should be completed before the first hurricane warning.

You must not wait until the forecasters tell you the hurricane is on its way.

You must have water.

You must have non-perishable foods.

You must have supplies for your baby.

You must have supplies for the elderly.

You must make arrangements for your pet/s.

You must have both prescription and over the counter meds.

You must know where you are going if you are told to evacuate.

You must take your kit with you if you are going to a shelter, most of them provide nothing else.

You must have your grill, coal/gas, matches, and old newspapers.

You must be able to meet your and your family's needs for at least (3) three days after the storm.

You must have "fun things for the kids." plus snacks.

You must fill your bathtub, (don't depend on the stopper hanging on that chain over the tub, the water might drain out if the cork is not tight enough, instead, buy a suction stopper at the dollar store, or anywhere else). Fill all the pails and pans you have in the house as well. If the water goes off, and it might, you will be set.

And don't forget America's most frequent disaster, fire, take care in your everyday life to avoid same.

-Watch for open flames,

-don't let the children play with matches and lighters,

-don't leave candles unattended,

- don't overload sockets,

Make sure the smoke detectors in your home work.

AND IN CONCLUSION

I hope this book will help in your preparation for Hurricanes or any other Disaster we might face in the future. Because of my experience with Hurricane Wilma, when there was no power in my neighborhood for over two weeks, I became painfully aware of how ill-prepared I was and became aware of all the necessities that were lacking in my, as well as friends and neighbors household.

Don't' forget to guard against fires as well and have a fire drill in your household twice per year with your children.

In the off seasons store your non-perishable supplies in a well organized closed so you can see what need replacing in a flash, and remember, you will always need batteries.

So again,

- listen to the forecasts, and
- begin shopping early
- stock up on all needed supplies and necessities.

You must try to make your life as comfortable as you can even without power. We cannot be certain how long we will be without power and how early we might get help. So remember to prepare for **before, during, and after.**

Remember these are hurricane/disaster recipes do your best to add your own special touches using what's available to you.

Remember get your check list long before the hurricane and make sure you have ***'every little thing'*** you might need.

Pick up extra cans when you go food shopping each week and you will not be overwhelmed when the hurricane is near.

Don't forget to put away a few dollars each payday and keep it at home, preferably with your Kit that it's easily assessable.

Safeguard your home.

Take pictures of your furniture and other valuables for insurance purposes.

Include the kids, pets and elderly in your plannings.

Good Luck and arrange to be safe.

Made in the USA
Monee, IL
04 November 2026

67378639R00049